YOUR KNOWLEDGE HAS VALUE

- We will publish your bachelor's and master's thesis, essays and papers

- Your own eBook and book - sold worldwide in all relevant shops

- Earn money with each sale

Upload your text at www.GRIN.com and publish for free

Bibliographic information published by the German National Library:

The German National Library lists this publication in the National Bibliography;
detailed bibliographic data are available on the Internet at http://dnb.dnb.de .

Imprint:

Copyright © 2011 GRIN Verlag, Open Publishing GmbH
Print and binding: Books on Demand GmbH, Norderstedt Germany
ISBN: 9783668575707

This book at GRIN:

http://www.grin.com/en/e-book/379556/construction-and-testing-of-single-and-
double-ring-infiltrometers

Ajayi Stanley, E.C. Eriakha

Construction and Testing of Single and Double Ring Infiltrometers

GRIN Publishing

CONSTRUCTION AND TESTING OF A SINGLE AND DOUBLE RING INFILTROMETERS IN AUCHI POLYTECHNIC

By

Eriakha E. C. and Ajayi A. S.

Department of Agricultural and Bio-Environmental Engineering Technology, Auchi Polytechnic Auchi

ABSTRACT

This paper presents the results of the construction of an improvised double and single ring infiltrometers. The study was carried in Auchi Polytechnic campus, after the construction, the double ring and single ring infiltrometer were used to carryout infiltration tests at four different locations. The t-test was used to test check the performance of the double and single ring infiltrometers. The results of the soil properties showed that the soil texture was predominantly sandy loam and loam respectively. The result presented herein showed a high significant difference between the performance of the single and double ring infiltrometer at Plot A and no significant other plots. The total cumulative infiltration depth of the single-ring infiltrometer for the A, B, and C and D fields were 188.30 cm, 59.70 cm, 77.00 cm and 97.00 cm, respectively. The total cumulative infiltration depth of the Double-ring infiltrometer for the A, B, and C and D fields were 123.30 cm, 75.60 cm, 70.50 cm and 95.20 cm, respectively. Generally, Philip's model performed better than Kostiakov's model

Key Words: Single and Double ring infiltrometer, infiltration models, Kostiakov,

Inhalt

1. INTRODUCTION

Infiltration is the process of water movement from the ground surface into the soil and is an important component in the hydrological cycle (Haghiabi *et. al.,* 2011). Infiltration characteristics of soils can be quantified by direct measurement on the field and/or when field infiltration data are fitted mathematically to infiltration models (Oku and Aiyelari, 2011). Lili *et al.,* (2008) reviewed the commonly used direct methods for measuring soil infiltration which include: single ring and double ring infiltrometers, mariotte-double ring infiltrometer, disc permeameter, rainfall simulator, runoff-on-ponding, runoff-on-out and linear source methods, the results obtained from field infiltration test and soil analysis are used for infiltration modeling.

The general knowledge of the infiltration rate of soils has been obtained for many field conditions in the past, and results applied in many field works, there is dearth of information on the infiltration characteristics of soils within Auchi Polytechnic campus which would have aided engineers in hydrologic/hydraulic design of water disposal structures, soil management and erosion control works et cetera. The knowledge of the infiltration capacity of soils within Auchi Polytechnic campus will be very helpful in carrying out soil related research works, infrastructural works and other works involving erosion control and management. This paper presents the results of the construction of an improvised double and single ring infiltrometers, carry out a comparative study of the variability of the cumulative infiltration depths of the single and double ring infiltrometers and also evaluate the ability of two infiltration models (Kostiakov's, and Philip's model) to predict cumulative infiltration for the five locations in comparison with the field obtained data.

2. MATERIALS AND METHODS
2.1. Study Area

This study was be carried out in specific locations within Auchi Polytechnic campus, Auchi Polytechnic is located between latitude 7^0 10' and 7^0 20' north of the equator and longitude 6^0 16' and 6^0 36' east of the Greenwich Meridian with an altitude of 207m. The locations chosen for infiltration runs are Agricultural & Bio-Environmental Engineering experimental field behind the workshop, School of ICT, School of Environmental studies, and Campus II, which were denoted by A, B, C and D respectively.

2.2 Construction Criteria

The development of an improvised double ring and single ring infiltrometer was constructed based on standard principles. This system involved the use of sheet metal of 2 mm thickness, this was chosen to avoid bending and so it can withstand the impact of the blow from the mallet while it is been driven into the soil. The outer ring of 50cm diameter and 40cm height with inner and single ring 30cm diameter and 40 cm height was constructed from the sheet metal. The meter rule, mallet, and other accompanying materials were purchased directly from the market.

2.3 Testing and Evaluation

After the construction, the double ring and single ring infiltrometer were used to carryout infiltration tests at four different locations. Undisturbed samples were also taken from each location at depth of $0 - 15$cm and $15 - 30$cm using core samplers to determine the soil's physical property of the study areas. Both rings were hammered 15 cm into the soil with a plank to protect the surface of the ring from damage during hammering. The test started by pouring water into the inner ring to an appropriate depth and at the same time, adding water to the space between the two rings to the same depth as quickly as possible. The time when the test began was recorded and the water level on the measuring rod was noted. After two (2) minutes, the drop in water level in the inner ring was recorded on the measuring rod and water added to bring the level back to approximately the original level at the start of the test. The water level outside the ring was maintained similar to the one inside. Each infiltration test lasted for about four (4) hours, the cumulative time intervals will be; 2, 4, 7, 10, 15, 20, 30, 45, 60, 80, 100, 120, 150, 180, 210 and 240 minutes. The cumulative infiltration depth at the elapsed time was being recorded accordingly.

2.4 Estimation of Model Parameters

In order to assess the performance of the selected models in predicting the cumulative infiltration, the parameters of each model were first determined as follows:

2.4.1. Kostikov's Equation

The Kostiakov's equation (1932), is given by:

$$I = kt^a \qquad \text{---------- [2.1]}$$

where: I = cumulative infiltration (cm), t= time from the start of infiltration (hr), and a and k are empirical parameters that need to be estimated.

When Eq. 2.1 is differentiated the infiltration rate i (cm/hr) will be obtained as:

$$i = akt^{a-1} \text{ ---------- } [2.2]$$

The parameters, k and a must be evaluated from measured infiltration data, since they have no physical interpretation.

The functional relationship between cumulative infiltration I, and time t, has been given by Eq. 2.1. The plot of I against t on a Log-Log graph gives k as the intercept of the graph and a as the slope. The values of a and k were now substituted into Eq. 2.1 and 2.2 at the different elapsed time to get the model generated cumulative infiltration and infiltration rates respectively.

2.4.2 Philip's Equation

Philip (1957) infiltration model is expressed as:

$$I = S\sqrt{t} + At \text{------------}[2.3]$$

The infiltration rate then becomes:

$$i = \frac{S}{2\sqrt{t}} + A \text{-------------}[2.4]$$

Where: S (cm/hr$^{1/2}$) is called Sorptivity and is a function of the boundary and initial water contents, θ_o, and θ_i.

The parameter A (cm/hr) is the Transmitivity or Permeability coefficient or gravity term which is equivalent to the saturated hydraulic conductivity of the soil.

The fitting parameters S and A for Philip's equation were evaluated and used to simulate cumulative infiltration and infiltration rate by obtaining a linear plot of the transformed cumulative infiltration It$^{-0.5}$ (cm/hr$^{-0.5}$) versus the transformed time t$^{-\frac{1}{2}}$ was plotted, the slope of the graph represents the parameter A and the intercept is the Sorptivity (S). S and A were then substituted into Eq. (2.3) and (2.4) to obtain the cumulative infiltration and infiltration rates respectively.

2.5 Statistical Analysis and Model Validation

A comparative assessment of the performance of single and double ring infiltrometers will be carried out with t-test ANOVA using Microsoft excel 2010. Also, In order to prove the performance of the models and their parameters, each model will be validated by comparing their simulated data with field measured data. From the two separate infiltration result obtained, one set of result was used for simulation and the second was used for model validation. The validation of the models will be done using: RMSE (root mean square error) and R^2 (coefficient of determination), RMSE values decreases with increasing precision. R^2 provides a measure of how well observed outcomes are replicated by the model; it ranges from 0 to 1.

$$R^2 = \frac{\sum_{i=1}^{n}(O_i - \tilde{O})^2}{\sum_{i=1}^{n}(P_i - \tilde{O})^2} \text{--Eq. 2.5}$$

$$RMSE = \sqrt{\frac{\sum_{i=1}^{n}(O_i - P_i)^2}{n}} \text{----------------------------------- Eq. 2.6}$$

Where: Pi = predicted values, $\tilde{O}$ = mean of the observed data, Oi = observed values, n = number of samples.

3. RESULT AND DISCUSSION

3.1 Soil Properties

The result of analysis of soil physical properties of the study area is presented in Table 3.1. The results showed that the texture of the field surface (0 - 15) cm and the sub-surface (15 - 30) cm depths for the three sampled strips were predominantly sandy loam and loam respectively according to the United States Department of Agriculture (USDA) classification.

Table: 3.1 Average soil physical characteristics of the strips

Location	Depth(cm)	%Sand	%Silt	%Clay	Textural Class	θ_i	θ_f	BD (g/cm^3)
A	0-15	60	22	18	Sandy Loam	12.6	44.1	1.48
	15-30	58	23	19	Sandy Loam	13.2	44.2	1.48
B	0-15	57	29	14	Sandy Loam	15.8	44.8	1.46
	15-30	61	20	19	Sandy Loam	13.2	43.9	1.49
C	0-15	45	34	21	Loam	14.3	45.4	1.45
	15-30	46	32	22	Loam	14.8	45.2	1.45
D	0-15	50	35	15	Loam	10.9	45.2	1.45
	15-30	56	29	15	Sandy Loam	10.9	44.8	1.46

*BD = Bulk density; θ_i = Initial Moisture content; θ_f = Initial Moisture content

3.2 Field Measured Infiltration

The results of the measured cumulative infiltration and calculated infiltration rate are summarized in Table 3.2 and 3.3.

Table: 3.2 Cumulative Infiltration

Time (hr)	A Single Ring	A Double Ring	B Single Ring	B Double Ring	C Single Ring	C Double Ring	D Single Ring	D Double Ring
	I (cm)	I (cm)	I (cm)	I (cm)	I (cm)	I (cm)	I (cm)	I (cm)
0.03	5.30	6.30	6.00	5.00	4.00	4.60	6.70	4.00
0.07	8.70	8.70	10.00	8.00	7.90	8.60	8.70	7.00
0.12	13.30	12.90	15.00	12.20	11.90	14.60	12.70	13.00
0.17	20.80	15.90	23.00	16.20	14.10	19.10	16.70	15.00
0.25	32.50	22.90	27.60	23.20	19.50	24.50	20.70	25.00
0.33	40.90	30.90	31.10	29.20	26.50	31.70	24.70	33.00
0.50	54.30	39.90	34.70	36.30	35.50	37.90	33.50	41.00
0.75	74.70	49.50	37.50	38.80	45.60	41.50	43.10	48.00
1.00	89.90	58.80	41.70	42.80	50.80	45.90	52.10	54.00
1.33	102.20	65.80	47.30	47.80	54.00	51.40	58.70	62.20
1.67	112.30	74.20	49.40	49.80	59.60	55.30	66.70	68.30
2.00	123.60	85.30	52.40	52.80	64.00	59.40	73.60	71.60
2.50	137.80	95.30	55.60	57.20	67.00	63.10	81.60	77.20
3.00	167.90	104.80	57.70	66.70	71.00	65.90	86.10	84.50
3.50	179.20	114.40	59.70	70.10	74.00	68.30	93.80	90.90
4.00	188.30	123.30	59.70	75.60	77.00	70.50	97.00	95.20

3.3 Comparison of the cumulative infiltration by the Single and Double ring Infiltrometers.

In order to access the variation in the performance of the single and double ring infiltrometers, t – test was conducted for the for locations under study, Table 3.4 presents the results of the calculated t-test at a confidence level of 0.05 with a critical values of 2.131 and degree of freedom 15.

Table 3.3 Result of T-test

	Plot A	Plot B	Plot C	Plot D
Number of Observations	16	16	16	16
Pearson Correlation	0.998	0.983	0.995	0.993
Degree of freedom	15	15	15	15
t Stat	4.727	-1.010	1.187	-0.871
t Critical	2.131	2.131	2.131	2.131

*If $t_{stat} > t_{crit}$ = There is a significant difference then if $t_{stat} > t_{crit}$ = There is no significant difference

The result above shows that there is a high significant difference between the performance of the single and double ring infiltrometer at Plot A which is the Agricultural engineering experimental plot and there is no significant difference between their performances for other plots. The total cumulative infiltration depth of the single-ring infiltrometer for the A, B, and C and D fields were 188.30 cm, 59.70 cm, 77.00 cm and 97.00 cm, respectively. The total cumulative infiltration depth of the Double-ring infiltrometer for the A, B, and C and D fields were 123.30 cm, 75.60 cm, 70.50 cm and 95.20 cm, respectively. While the total accumulated infiltration in the single-ring infiltrometer was about 35%, 9% and 2 % higher than that of the double-ring infiltrometer in the A, C and D fields, respectively, while the total accumulated infiltration in the double-ring infiltrometer was 20%. The result shows that the accumulated infiltrations of the fallowed fields (B, C and D) were generally lower than the cultivated field A. This may be attributed to the fact that the fallowed fields were covered with vegetation.

3.4 Model's Parameter Evaluation

The process of estimating the model's parameters and time exponent differs for each model, each of the models was first transformed into its linear equivalent in which to have dependent and independent variables respectively and the coefficients of the linear functions read from the graphs are the model's parameters. The following figures are the graphical relations to obtain the fitting parameters.

3.4.1 The Kostiakov's Model

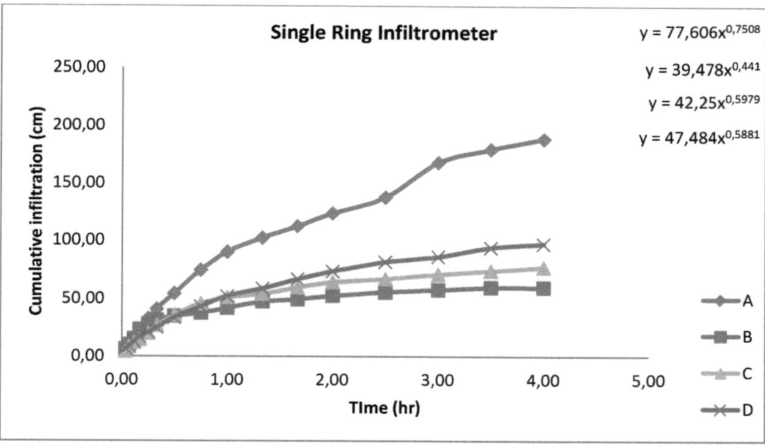

Figure 3.1: The graph of Cumulative infiltration versus elapsed time to obtain the fitting parameters for Kostiakov's equation for Single ring test.

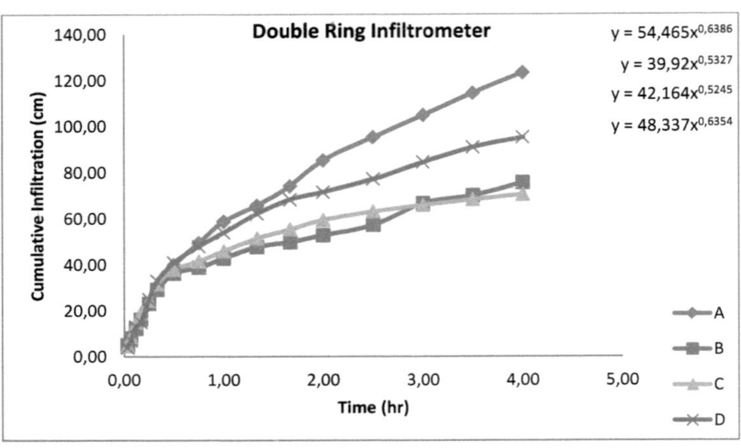

Figure 3.2: The graph of Cumulative infiltration versus elapsed time to obtain the fitting parameters for Kostiakov's equation for Double ring test.

Table: 3.4 Kostiakov model's parameters and modelled equations for Single ring test

Strip	Parameter values or estimated constant		Modelled equations
	k	a	
A	77.606	0.751	$I = 77.606t^{0.751}$
B	39.478	0.440	$I = 39.478t^{0.440}$
C	42.250	0.598	$I = 42.250t^{0.598}$
D	47.484	0.588	$I = 47.484t^{0.588}$

Table: 3.5 Kostiakov model's parameters and modelled equations for Double ring test

Strip	Parameter values or estimated constant		Modelled equations
	k	a	
A	54.465	0.639	$I = 54.465t^{0.639}$
B	39.920	0.533	$I = 39.920t^{0.533}$
C	42.164	0.525	$I = 42.164t^{0.525}$
D	48.337	0.635	$I = 48.337t^{0.635}$

3.4.1 Philip's model

The values of the parameters (*A and S*) of Philip's model are given below; the evaluation tables and graphs are presented in Appendix E

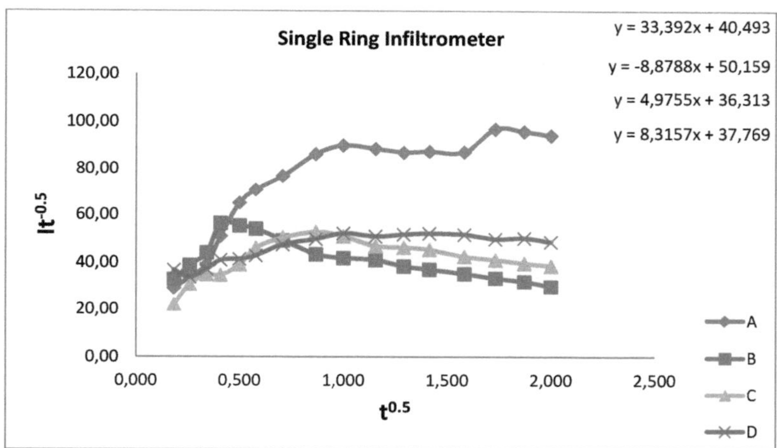

Figure 3.3: The graph of Transformed Cumulative infiltration versus transformed elapsed time to obtain the fitting parameters for Philips's equation for single ring test.

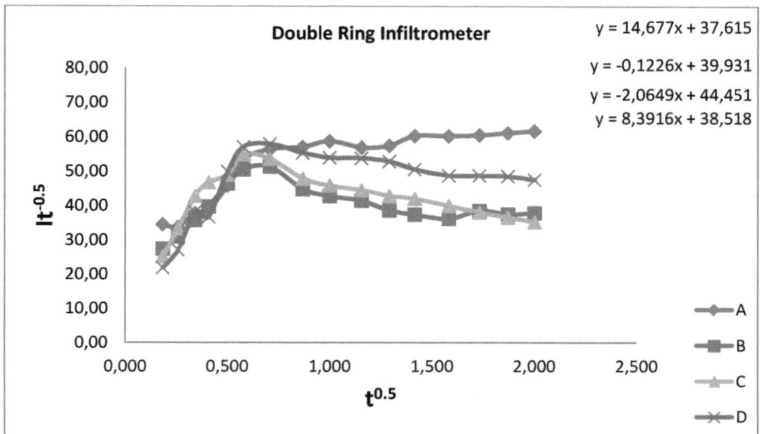

Figure 3.4: The graph of Transformed Cumulative infiltration versus transformed elapsed time to obtain the fitting parameters for Philips's equation for Double ring test.

Table: 3.6 Philip's model parameters and modelled equations for single ring test

Strip	Parameter values or estimated constant		Modelled equations
	S	A	
A	40.493	33.392	$I = 40.493\sqrt{t} + 33.392t$
B	50.159	-8.879	$I = 50.159\sqrt{t} - 8.879t$
C	36.313	4.976	$I = 36.313\sqrt{t} + 4.976t$
D	37.769	8.316	$I = 37.769\sqrt{t} + 8.316t$

Table: 3.7 Philip's model parameters and modelled equations for double ring test

Strip	Parameter values or estimated constant		Modelled equations
	S	A	
A	37.615	4.677	$I = 37.615\sqrt{t} + 4.677t$
B	39.931	-0.123	$I = 39.931\sqrt{t} - 0.123t$
C	44.451	-2.064	$I = 44.451\sqrt{t} - 2.064t$
D	38.518	8.392	$I = 38.518\sqrt{t} + 8.392t$

3.5 Simulation of Cumulative Infiltration using the Estimated Parameters

The values of the parameters estimated shown in Tables 4.1 to 4.4 were then incorporated into the two infiltration models and simulation of cumulative infiltration was made for each of the fields and the predicted cumulative infiltration were compared with the measured cumulative infiltration with R^2 and RMSE.

3.6 Model Validation

Tables 4.5 and 4.6 also shows the statistical indices of the comparison between the model simulated and observed infiltration for the CM, PM and Control strip respectively.

Table: 3.8 Observed and Model predicted cumulative infiltration for Single ring test

Time	Field A			Field B		
(hr)	I (cm)	Kost	PH	I(cm)	Kost	PH
0.03	5.30	6.04	8.51	6.00	8.81	8.86
0.07	8.70	10.16	12.68	10.00	11.96	12.36
0.12	13.30	15.47	17.73	15.00	15.31	16.10
0.17	20.80	20.21	22.10	23.00	17.91	19.00
0.25	32.50	27.41	28.59	27.60	21.42	22.86

0.33	40.90	34.02	34.51	31.10	24.32	26.00
0.50	54.30	46.12	45.33	34.70	29.08	31.03
0.75	74.70	62.53	60.11	37.50	34.77	36.78
1.00	89.90	77.61	73.89	41.70	39.48	41.28
1.33	102.20	96.32	91.28	47.30	44.82	46.08
1.67	112.30	113.88	107.93	49.40	49.45	49.96
2.00	123.60	130.59	124.05	52.40	53.59	53.18
2.50	137.80	154.41	147.51	55.60	59.14	57.11
3.00	167.90	177.06	170.31	57.70	64.09	60.24
3.50	179.20	198.78	192.63	59.70	68.59	62.76
4.00	188.30	219.75	214.55	59.70	72.76	64.80
R^2	**0.984**	**0.979**		**0.945**	**0.976**	
RMSE	**11.899**	**10.492**		**5.456**	**2.960**	

Field C			**Field D**		
I(cm)	**Kost**	**PH**	**I(cm)**	**Kost**	**PH**
4.00	5.53	6.80	6.70	6.42	7.17
7.90	8.37	9.71	8.70	9.66	10.31
11.90	11.69	12.98	12.70	13.42	13.87
14.10	14.47	15.65	16.70	16.55	16.81
19.50	18.44	19.40	20.70	21.01	20.96
26.50	21.91	22.62	24.70	24.89	24.58
35.50	27.92	28.16	33.50	31.59	30.86
45.60	35.57	35.18	43.10	40.09	38.95
50.80	42.25	41.29	52.10	47.48	46.08
54.00	50.18	48.56	58.70	56.24	54.70
59.60	57.34	55.17	66.70	64.12	62.62
64.00	63.95	61.31	73.60	71.38	70.04
67.00	73.07	69.85	81.60	81.39	80.51
71.00	81.49	77.82	86.10	90.60	90.36
74.00	89.36	85.35	93.80	99.20	99.76
77.00	96.78	92.53	97.00	107.30	108.80
	0.942	**0.941**		**0.990**	**0.984**
	8.101	**6.908**		**3.614**	**4.360**

Table: 3.9 Observed and Model predicted cumulative infiltration for Double ring test

Time (hr)	Field A			Field B		
	I (cm)	Kost	PH	I(cm)	Kost	PH
0.03	6.30	6.21	7.36	5.00	6.52	7.29
0.07	8.70	9.66	10.69	8.00	9.43	10.30
0.12	12.90	13.81	14.56	12.20	12.71	13.62
0.17	15.90	17.35	17.80	16.20	15.37	16.28
0.25	22.90	22.47	22.48	23.20	19.08	19.93
0.33	30.90	27.00	26.61	29.20	22.23	23.01
0.50	39.90	34.98	33.94	36.30	27.60	28.17
0.75	49.50	45.32	43.58	38.80	34.25	34.49
1.00	58.80	54.47	52.29	42.80	39.92	39.81
1.33	65.80	65.45	63.00	47.80	46.53	45.94
1.67	74.20	75.47	73.02	49.80	52.40	51.35
2.00	85.30	84.79	82.55	52.80	57.75	56.23
2.50	95.30	97.78	96.17	57.20	65.04	62.83
3.00	104.80	109.85	109.18	66.70	71.67	68.79
3.50	114.40	121.22	121.74	70.10	77.81	74.27
4.00	123.30	132.01	133.94	75.60	83.54	79.37
	R^2	0.994	0.989		0.970	0.975
	RMSE	3.837	4.655		5.106	3.876

Field C			Field D		
I(cm)	Kost	PH	I(cm)	Kost	PH
4.60	7.08	8.05	4.00	5.57	7.31
8.60	10.19	11.34	7.00	8.65	10.50
14.60	13.66	14.94	13.00	12.34	14.14
19.10	16.47	17.80	15.00	15.48	17.12
24.50	20.38	21.71	25.00	20.03	21.36
31.70	23.70	24.98	33.00	24.05	25.04
37.90	29.31	30.40	41.00	31.12	31.43
41.50	36.26	36.95	48.00	40.26	39.65

45.90	42.16	42.39	54.00	48.34	46.91
51.40	49.03	48.57	62.20	58.03	55.67
55.30	55.12	53.94	68.30	66.87	63.71
59.40	60.65	58.73	71.60	75.09	71.26
63.10	68.18	65.12	77.20	86.52	81.88
65.90	75.02	70.80	84.50	97.15	91.89
68.30	81.34	75.93	90.90	107.15	101.43
70.50	87.24	80.64	95.20	116.63	110.60
	0.948	**0.963**		**0.961**	**0.960**
	6.979	**4.763**		**9.021**	**7.097**

The coefficients of determination (R^2) between the field-measured and model simulated data were very high (> 0.94) which implied that the models were able to simulate water infiltration in the study areas adequately. The result of the coefficient of determination (R^2) ranged from 0.941 to 0.994 which are all close to unity and an indication of close agreement between the measured and predicted data for each of the infiltration models.

Considering the individual performance of the models in the areas, for the sing ring test, the Kostiakov model performed better than Philip's model in field A, C and D while the Philip's model predicted better in field B, the for the double ring test the Kostiakov's model performed better in field A and D while the Philip's model performed better than Kostiakov in field B and C with R^2 values in the tables above. The result above proves that both models are efficient in simulating water movement into the soil.

In order to further check the discrepancies between the predicted and the measured values, Root Mean Square Error (RMSE) was used. The closer the prediction is to zero the better the model, the RMSE values ranged from 2.960 – 11.898 for the study areas. The result shows that the Kostiakov's model had the largest value of RMSE, for the single ring test, Philip's model minimized errors of prediction in field A, B and C while the Kostiakov's model did better only in field D. then for the double ring test, Kostiakov's model minimized the error only in field A, while Philip's model performed better in field B, C and D.

We can therefore conclude that Philip's model performed better than Kostiakov's model, this is contrary to the work by Igbadun and Idris (2007), who observed that Kostiakov (1932) model, fitted experimental data better than Philip (1957) model for a hydromorphic soil at Samaru, Nigeria. The result of this study agrees with the findings of Al-Azawi (1985), who evaluated six infiltration models on a relatively homogenous, coarse-textured soil. He found that Philip's model gave a very good representation of the infiltration while than Kostiakov.

4. CONCLUSION

The design, fabrication and performance evaluation of a single ring and double ring infiltrometer was done in this work. The total cumulative infiltration depth of the single-ring infiltrometer for the A, B, and C and D fields were 188.30 cm, 59.70 cm, 77.00 cm and 97.00 cm, respectively. The total cumulative infiltration depth of the Double-ring infiltrometer for the A, B, and C and D fields were 123.30 cm, 75.60 cm, 70.50 cm and 95.20 cm, respectively. Similar work can be carried out to study the effect of different soil management practice (mulched land, tilled land, compacted soil, cultivated field e.t.c, on the infiltration capacity and soil physical properties of the areas studied.

REFERENCES

Al-Azawi, S. A. (1985). Experimental evaluation of infiltration models. *Journal of Hydrology. 24(2): 77 – 88.*

Haghiabi, A. H., Heidarpourand M. and Habili, J. (2011). A new method for estimating the parameters of Kostiakov and modified Kostiakov infiltration equations. *World Applied Sciences Journal.* 15(1): 129 –135.

Igbadun, H. E. and Idris, U. D. (2007). Performance Evaluation of Infiltration Models in a Hydromorphic Soil. *Nig. Journal of Soil &Env.* Res. Vol. 7: 2007, 53-59.

Kostiakov, A. N. (1932). On the dynamics of the coefficient of water-percolation in soils and on the necessity for studying it from a dynamic point of view for purposes of amelioration. *Transactions Congress International Society for Soil Science, 6th, Moscow, Part A: 17-21.*

Lili, M., Bralts, V. F., Yinghua, P., Han, L. and Tingwu L. (2008). Methods for measuring soil infiltration. State of the art, *Int. J.Agric&Biol Eng.* 1(1), 22-30.

Oku, E. and Aiyelari, A. (2011). Predictability of Philip and Kostiakov infiltration model under inceptisols in the Humid Forest Zone, Nigeria. *Kasetsart Journal (Natural Science), 45: 594 -602.*

Philip, J. R. (1957). The theory of infiltration: 1. The infiltration equation and its solution *Soil Science.83, 345-357.*

YOUR KNOWLEDGE HAS VALUE

- We will publish your bachelor's and
  master's thesis, essays and papers

- Your own eBook and book -
  sold worldwide in all relevant shops

- Earn money with each sale

Upload your text at www.GRIN.com
and publish for free